悦读

A VISUAL FEAST

百年上理优秀
历史保护建筑

Architectural Heritage
of Centennial USST

吴坚勇　丁晓东 主编

Editors-in-Chief
Wu Jianyong & Ding Xiaodong

全国重点文物保护单位
**Key Historical and Cultural Site
under National Protection**

上海书画出版社
Shanghai Fine Arts Publisher

每一艘开往上海的轮船都必须在这所大学的视线内经过；
在这样一个校园里，任何有思想的学生都不能不感到自己生活在一个大的世界里。

EVERY SHIP THAT GOES TO SHANGHAI MUST PASS WITHIN FULL VIEW
OF THE COLLEGE; ANY THOUGHTFUL STUDENT ON SUCH A CAMPUS IS
COMPELLED TO LIVE IN A LARGE WORLD.

浦江之滨，花木扶疏，红楼三五，矗立其间，上理在焉。

上海理工大学素有"沪上最美校园"之称，承续于 1906 年创办的沪江大学和 1907 年创办的德文医工学堂。

典雅的军工路主校区完整保留了沪江大学时期建成的 36 幢哥特式风格建筑，是上海高校中规模最大的市级优秀历史建筑群，2019 年被国务院公布为"全国重点文物保护单位"。精致的复兴路校区矗立着 7 幢德文医工学堂时期建成的普鲁士风格建筑，其中的工科讲堂被誉为"在中国的普鲁士风格建筑的标志"。

建筑是古老的艺术，是历史的印记，是空间的时代意志，是人类创造的奇迹。有什么样的建筑，就有什么样的大学。建筑与大学互为表达。阅读这些建筑，就是阅读我们的大学。

我们阅读学校第一座大建筑思晏堂，聆听陈独秀、孙中山呼吁青年学子投身革命的激情演说；阅读"沪上各校之冠"的工科讲堂，自德国和法国远洋运来的实验设备高端精良；阅读民国诗人徐志摩捐建的麦氏医院，和他在诗意的校园里徜徉沉思；阅读"沪上各校第一座"体育馆，体操队、篮球队、网球队、足球队击败华东各校纷纷夺冠；阅读"国内仅见之建筑物"的科学馆，从中走出纪育沣、涂长望等 18 位两院院士；阅读由美国著名建筑师墨菲设计的图书馆，设施先进、用升降机传递图书……

这些建筑见证了中国大学教育发展历程。20 世纪初，世界风云变幻，中国处于历史转折关头，晚清政府废科举鼓励兴办新式学堂，始有沪江大学与德文医工学堂的创建。这些建筑具有独特文化价值和审美价值，荷载着学校的精神传统。我们阅读美轮美奂的建筑空间，更是阅读上理中西合璧、典雅优美的人文校园，阅读上理扎根中国、面向世界的办学理念，阅读上理海纳百川、兼容并蓄的文化品格，阅读上理爱党爱国、传承不息的红色基因。

2018 年，校长丁晓东提出"校园可漫步，建筑可阅读，文化可触摸"，希望讲好学校历史建筑故事，让师生"可阅读"。档案馆广泛搜集档案史料，在已有研究基础上，对历史建筑的建造时间、人文故事等内容进行源头考证，设计二维码铭牌，录制音频，实现了历史建筑的可证、可读、可听、可感和可思，凸显文化内涵。

为进一步落实李强书记"建筑可阅读"文旅融合工作的指示，校党委书记吴坚勇、校长丁晓东启动并主持《悦读：百年上理优秀历史保护建筑》画册的编辑出版工作。在校党委宣传部的支持下，档案馆制作建筑画册。画册图文并茂、中英对照，展现了上理优秀历史建筑风貌，勾勒出百十年发展轨迹，彰显了学校深厚的历史文化底蕴。

习近平总书记指出"要让文物说话，让历史说话，让文化说话"（《习近平关于社会主义文化建设论述摘编》，中共中央党史和文献研究院编，中央文献出版社，2017 年 10 月）。黄浦江畔巍巍学府，沧桑百年薪火相传。上理在百十年的办学实践中形成了深厚的文化积淀和光荣的革命传统，一代代上理人接续奋斗，为祖国、为人民、为民族做出了杰出贡献。希望透过《悦读：百年上理优秀历史保护建筑》画册，漫步校园，阅读上理，触摸历史，在波澜壮阔的新时代中感受大哉上理！

PREFACE

出版前言

CONTENT

出版前言

Alongside the Huangpu River stands USST (the University of Shanghai for Science and Technology) amid the verdure and blossoms of its campus, highlighted by its preserved historical buildings made of red bricks here and there.

With a long-standing reputation as "the most beautiful campus in Shanghai", USST originated from the merger of the University of Shanghai (founded in 1906) and the Deutsche Medizinschule (established in 1907).

Boasting 36 intact Gothic buildings built during the period of the University of Shanghai, the Jungong Road Campus of USST, which was listed as a "Key Historical and Cultural Site under National Protection" by the State Council in 2019, constitutes the largest group of university buildings of outstanding historical importance in the city. Seven Prussian-style buildings erected in the period of the Deutsche Medizinschule stand on its Fuxing Road Campus, of which the Engineering Courses Hall is acclaimed as a "Prussian-style landmark in China".

Architecture serves as an antique art, an imprint of a historical moment and a testament to the use of space by past generations: it is, therefore, a miracle created by humans. Architecture reflects the university. They speak for each other. To read these buildings is to read our university.

Reading the campus architecture, we read the university's history and culture. We understand Yates Hall, the first large construction in the university, as in it Chen Duxiu and Sun Yat-sen made passionate speeches calling upon the young students to devote themselves to the revolution; we see the Engineering Courses Hall, the best edifice belonging to universities in Shanghai, and we envision the advanced and elaborate lab equipment imported from Germany and France; we view the Mcleish Infirmary, donated by Xu Zhimo(also known as Hsu Chih-mo), a renowned poet during the Republic of China era (1912-1949), to envisage him wandering around the idyllic campus, contemplating its mysteries and truths; we read Haskell Gymnasium, the first gymnasium built for a college in Shanghai, to witness our triumphs over other local universities in gymnastics, basketball, tennis and football; we look at the Science Hall, one of the most advanced and pioneering buildings in the Republic of China era, and remember the 18 academicians, including Ji Yufeng and Tu Changwang, who have studied and worked there; we also

PREFACE
出版前言

gaze at the Library, designed by Henry Killam Murphy, the prestigious American architect, to experience its state-of-the-art technology, where books are retrieved and delivered by elevator....

These buildings have witnessed the developments of China's higher education. At the beginning of the 20th century, the world was changing and China was at a turning point in history: The late Qing government abolished the imperial examination and encouraged the establishment of new-style schools, which led to the founding of the University of Shanghai and the Deutsche Medizinschule. These buildings possess unique value in cultural and aesthetic senses, and bear the spirit and soul of the university. To look at the gorgeous buildings of USST is to relish its elegant, charming, cultural campus, which integrates the East and the West into a harmonious whole, to admire its school-running philosophy rooted in China while facing the world, to dig its inclusive culture, and to understand its everlasting red gene for its love for the Party and the country.

In 2018, Ding Xiaodong, President of USST, put forward the idea of "a walkable campus, readable buildings and a touchable culture", and called for well-told and readable accounts of the university's historical buildings for teachers and students. Based on existing research, the Archives has made the historical buildings verifiable, readable, audible, perceptible and thought-provoking by collecting a wide range of archival sources and historical files, conducting source textual research on the construction time and the cultural context of each historical building, designing corresponding QR code nameplates and recording audios to highlight their cultural significance.

In order to further implement Secretary Li Qiang's idea of readable buildings to integrate culture with tourism, Wu Jianyong, Party Committee Secretary of USST, and Ding Xiaodong, President of USST, launched the compilation and publication of the collection of paintings named A Visual Feast: Architectural Heritage of Centennial USST. With the support of the school's Party Committee's Publicity Department, the Archives has produced this architectural album with colorful pictures, well-annotated in both Chinese and English, fully demonstrating the style and features of the excellent historical buildings of USST, depicting its development over the decades, and reflecting its rich and abundant historical and cultural heritages.

PREFACE
出版前言

Xi Jinping, General Secretary of CPC, once pointed out, "Let cultural relics speak, let history speak and let culture speak." (*Excerpts from Xi Jinping's Discourse on Socialist Cultural Construction*, edited by Central Party History and Documentation Research Institute, Central Party Literature Press, October, 2017) Alongside the Huangpu River stands USST majestically; over the years, its traditions and ethos do not cease or change. Over the past century, USST has developed a profound culture and formed a glorious revolutionary tradition in its school-running experience. Generations of USST's teachers and students have spared no effort to make outstanding contributions to the motherland, the people and the nation. It is hoped that *A Visual Feast: Architectural Heritage of Centennial USST* can take you to walk around the campus, read the buildings and touch the history of the great USST in the New Era!

PREFACE
出版前言

CONTENT

目录

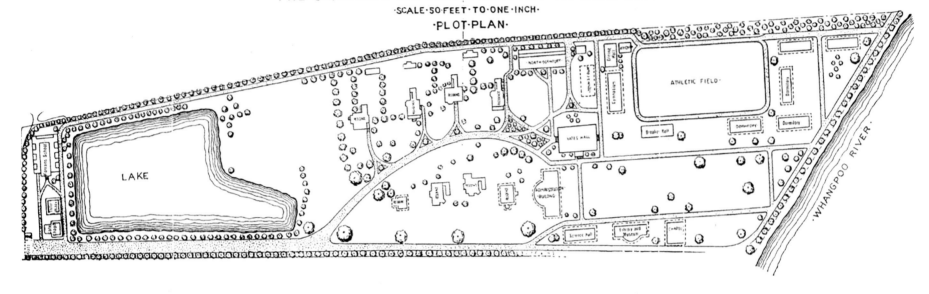

1907 年沪江大学的创建者初步规划了校园。

1919 年美国著名建筑设计师亨利·墨菲对沪江校园进行整体设计规划。

墨菲还规划设计了清华大学、金陵女子大学、燕京大学等多所大学校园。

—

Back in 1907, the founders of the University of Shanghai made an initial plan of the campus.

In 1919, Henry Killam Murphy (1877-1954), a famous American architect, who also planned and designed the

campus of Tsinghua University, Ginling College, Yenching University and other universities,

created an overall designing plan of the campus.

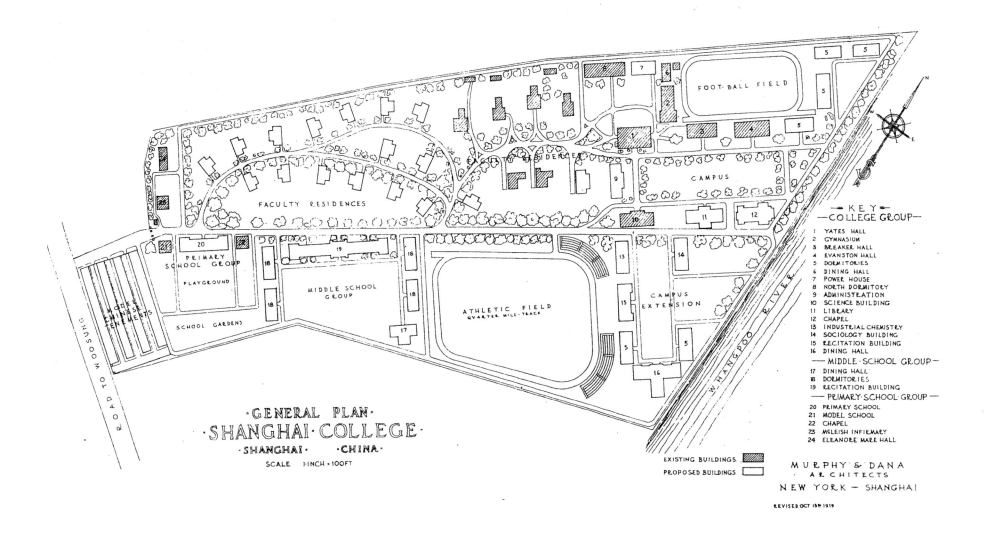

·GENERAL PLAN·
·SHANGHAI·COLLEGE·
·SHANGHAI· ·CHINA·
SCALE 1-INCH = 100 FT

→KEY←
—COLLEGE GROUP—
1 YATES HALL
2 GYMNASIUM
3 BREAKER HALL
4 EVANSTON HALL
5 DORMITORIES
6 DINING HALL
7 POWER HOUSE
8 NORTH DORMITORY
9 ADMINISTRATION
10 SCIENCE BUILDING
11 LIBRARY
12 CHAPEL
13 INDUSTRIAL CHEMISTRY
14 SOCIOLOGY BUILDING
15 RECITATION BUILDING
16 DINING HALL
—MIDDLE·SCHOOL·GROUP—
17 DINING HALL
18 DORMITORIES
19 RECITATION BUILDING
—PRIMARY·SCHOOL·GROUP—
20 PRIMARY SCHOOL
21 MODEL SCHOOL
22 CHAPEL
23 McLEISH INFIRMARY
24 ELEANORE MARE HALL

EXISTING BUILDINGS ▨
PROPOSED BUILDINGS ☐

MURPHY & DANA
·ARCHITECTS·
NEW YORK — SHANGHAI
REVISED OCT 15th 1919

墨菲沿用弗吉尼亚大学校园模式，将沪江校园分为大学区、中学区、小学区和教师住宅区，
主建筑群形成三面围合、中央绿地草坪的三合院布局。校园与江水相映、草地与树林相间，风光秀美，
建筑典雅，处处体现自由和谐、开放包容和注重人格培养的教育理念。

—

Adopting the campus style and design of the University of Virginia, Murphy divided the then Shanghai
College Campus into four separate zones – the college zone, the middle school zone, the elementary school zone and the
faculty residential zone. The major building group is enclosed within a three- sided courtyard with a green lawn as its center.
Alongside the Huangpu River amid lush lawns and abundant groves, the campus echoes with water,
with gorgeous landscape and elegant buildings, fully reflecting its educational philosophy of freedom,
harmony, openness, inclusiveness and emphasis on character cultivation.

军工路校区

JUNGONG ROAD CAMPUS

复兴路校区

FUXING ROAD CAMPUS

1905 年，科举制诏令废除。两个月后，沪江大学（以下简称"沪大"）董事会选定并购入了黄浦江西侧一块荒芜的江滩地。他们一眼便选中了这里，因为"这块地有一个优点，就是对黄浦江这条上海大都市与世界商业往来的通道一览无余"。

–

In 1905, the imperial examination system was abolished by governmental decrees. Two months later, the Board of Trustees of the University of Shanghai decided on the purchase of a piece of barren marshland on the west side of the Huangpu River. They chose to buy the site at first sight, claiming that "this plot is superior in that it commands a panoramic view of the Huangpu River--the trading passage between the metropolitan city of Shanghai and the world".

–

其后三十多年间，思晏堂、科学馆等 40 余幢哥特式风格建筑相继落成。沪大逐渐从"在中国的大学"变为"中国的大学"，以"学术化、人格化、平民化、职业化"为蓝图，以"提供学术领袖，培养为社会服务并使自己和周围人具有良好生活品质的人"为育人目标，成为当时国内一流大学。

–

Over the next three decades, more than 40 Gothic buildings, including Yates Hall and Haskell Gymnasium, were put up, one at a time. The University of Shanghai had gradually changed from "a university in China" to "a China's university", with its vision of "being academic, personalized, civilian and professional", and with its educational goal of "creating academic leaders and cultivating those who can serve society and improve the quality of life for both themselves and the people around them". By following this ethos, the University of Shanghai soon grew into a leading university in China in this period.

–

沪大坚持学术性与应用性并重，化学、社会学、教育学、生物学和工商管理学等国内一流，政治学、地质学和音乐等学科颇有建树。20 世纪 30 年代，学校以文风淳朴、较少教会气和爱国主义传统闻名全国，以文、理、商三科著称于世，办学规模和社会影响力与日俱增。

–

The University of Shanghai adhered to the philosophy of delivering both academic and applied courses, with such domestically first-class courses as chemistry, sociology, pedagogy, biology and business management, and such domestically outstanding theoretical courses as politics, geology, and music. In the 1930s, renowned for its disciplines of liberal arts, basic sciences and business, the university gained a nationwide reputation for its unsophisticated ethos, its less cloistered academic atmosphere and its patriotic tradition, witnessing an increase in both its scale and social influence.

PART 1

思晏堂

沪大第一座大建筑，见证学校百年发展变迁。
1907 年耗时 8 个月垫高地基，1908 年举行奠基礼并于年底竣工。
砖木结构，长方形平面，清水红砖砌筑，双坡红瓦屋面，南向门廊用组合柱式，东面入口为拱形。
建筑上的双联尖券、山墙玫瑰窗、尖塔和扶壁均具显著的哥特式特征。
1956 年遭龙卷风损坏，1957 年重修。

—

As the first large building of the University of Shanghai campus, Yates Hall has witnessed the developments and vicissitudes of the university. The foundation was dug for eight months in 1907, and the corner stone was laid in 1908. The Hall was erected at the end of the year. It consists of the brick-timber structure, rectangular plane, red brick masonry without plastering, double pitched red tile roof, south porch with composite columns, and an arched entrance to the east. The double pointed arch, rose windows, towers and buttresses, all feature a distinctive Gothic style. The Hall was seriously damaged by a tornado in 1956 and repaired in 1957.

YATES HALL

茂林深嶹，嘉树当楸。

1920 年 3 月 24 日，陈独秀在思晏堂演讲《什么是"新文化运动"？》；
5 月 25 日，孙中山在此演讲《中国之再造》。
徐志摩、李一氓、倪征噢、李公朴等历史文化名人都曾在思晏堂读书学习。
著名法学家吴经熊回忆："有一天，我在化学试验室做氧气试验时……好奇心大发，
想看看氧在瓶里会怎样燃烧。我试着用火柴点燃它，但瓶子马上就爆成了碎片。
当时我凑得很近以便于观察，却万幸未受损伤。"

—

On March 24, 1920, Chen Duxiu(1879-1942), delivered a speech entitled *What is the New Culture Movement?*
On May 25, Sun Yat-sen (1866-1925), addressed the students with a speech entitled *The Reproduction of China.*
Other significant figures in history such as Xu Zhimo, Li Yimeng, Ni Zhengyu and Li Gongpu, all studied
here.Wu Jingxiong, a well-known jurist, explained that: "When I was doing an experiment with oxygen in the
chemistry lab… curiosity occurred in my mind about how oxygen would burn in a bottle.
I tried to light it with a match, but the bottle immediately exploded into pieces when I was approaching closer
to observe the experiment; fortunately, I was safe and sound."

思晏堂晚秋

郑章成

烟水伶汀入画图，天高万叠乱云铺。

群山逶迤分浓淡，对树模糊认有无。

半晴天明远一坡，行火雁落平湖。

渔吹吗罢人何处，月映寒汀水绕芦。

万里云罗淡未收，坐火鸦藏获月秋。

渔灯暗逗芦村远，藏火箱在江楼。

烟波翠微归一鹤，潮平沙阔泛双鸥。

晚来似觉嫌云密，未听银蟾伴夜游。

晚明雾色爱晴霞，水郭荒村望眼赊。

对岸绿林古酒市，隔江红楚卖鲈家。

草枯驿路秋嘶马，风冷霜天幕集鸦。

时有野翁遥唱返，肩携黄菊一铡花。

郑章成，沪江大学第一届毕业生，曾任沪江书院院长、理学院院长、生物系主任，在沪江大学读书时赋诗一首：

1908 年 1 月 21 日，学校举行思晏堂奠基礼。
美国南北浸会第一届联席会议主席斯蒂芬斯为思晏堂奠基，
美国驻沪总领事田夏礼和上海道台梁如浩出席观礼。
大批客人从上海坐轮船沿黄浦江而下，登上江边泥滩。

魏馥兰校长记载：
"在沼泽芦苇丛中有一大堆土，一片脚手架，一堆堆准备起楼的砖，正中间是
一方石头……但是，这石头不仅是思晏堂的奠基石，也是整个学校的奠基石。"
—

On January 21, 1908, the university held the corner-stone laying ceremony of Yates
Hall. The Hon. E. W. Stephens, President of the Joint Convention of Northern and
Southern Baptists of America, laid the corner stone. Charles Denby, Jr., Consul-
General of the United States in Shanghai, and Liang Ruhao, circuit governor (superior
to a prefectural governor in status but inferior to a provincial governor) of Shanghai
were in attendance. Numerous guests came over from Shanghai in a steam boat,
landing with great difficulty on the muddy shore of the Huangpu River.

According to President Francis Johnstone White:
"A pile of mud in the midst of a reed marsh, a rack of scaffolding, piles of bricks
ready for the walls, and a cube of stone in the midst. This was the corner stone not
only of Yates Hall, but of the whole institution."

YATES
HALL

图书馆

1928 年 2 月 25 日，沪大首任华人校长刘湛恩博士就职典礼后，
举行图书馆破土礼。同年 9 月竣工，耗费 4 万美元。
1948 年图书馆向东扩建，次年启用。命名为"湛恩纪念图书馆"，以纪念为国殉难的刘湛恩校长。

—

On February 25, 1928, Dr. Liu Zhan'en (Herman C. E. Liu), the first Chinese president of the
University of Shanghai, held a groundbreaking ceremony after his inauguration ceremony. In September,
the Library was erected at a total cost of US $ 40,000. It was expanded eastwards in 1948, and put into use in 1949.
The library was named "Zhan'en Memorial Library" in honor of Liu Zhan'en, the first Chinese president of the
university, who laid down his life for the honor of the country.

LIBRARY

图书室原在思晏堂二楼。图书馆落成后，
学生排着长队，将思晏堂的图书一本一本传过来。
校西乐队在一旁奏乐鼓劲，学生们和着琴声唱歌，
欢欣雀跃，其乐融融。

—

The Reading Room was originally located on the second floor of Yates Hall. After the completion of the Library, students formed in a long human chain, passing books from Yates Hall one after another to the Library. When the books were being passed along the human chain, the school band played music while all the students cheered and sang along to the music, creating a hilarious and harmonious scene.

1928 年 11 月 17 日，图书馆举行开馆典礼。
民国四大书法家之首的谭延闿题写牌匾，
"图书馆"三字拙朴雄浑。来宾云集，外交部部长王正廷、
上海市教育局局长韦悫、中国公学校长胡适、东方图书馆
馆长王云五等社会名流出席。
胡适、王正廷、王云五发表了精彩演说。

—

On November 17, 1928, the university held the opening ceremony of the Library. Tan Yankai, the most famous figure of the Four Great Calligraphers of the Republic of China era, donated a plaque to the Library, on which the three simple, graceful, vigorous Chinese characters "图书馆" (library) were written. On this occasion, a myriad of guests was present, including such prominent social figures as Wang Zhengting, Minister of the Ministry of Foreign Affairs, Wei Que, Director of the Bureau of Education of Shanghai, Hu Shi, President of Wusong Public School, and Wang Yunwu, Curator of the Oriental Library. Hu Shi, Wang Zhengting, and Wang Yunwu delivered brilliant speeches.

LIBRARY

图书馆由美国设计师墨菲设计，共两层，支架为全钢骨结构。
扩建后面积达 2263 平方米，华美壮丽，气势恢弘。馆内设施先进，用升降机传递图书，方便快捷。
软木地板上铺有油地毡，墙壁四周安装暖气片。

—

Designed by the American architect Henry Murphy, the Library, built with a support structure of reinforced steel,
consists of two floors. After expansion, its area covers 2,263 square meters and constitutes a sumptuous and magnificent space.
Equipped with advanced facilities, the library made use of an elevator to pass books between floors.
The cork floor was covered with linoleum and the walls were equipped with radiators.

科学馆

1921 年始建，1922 年 4 月竣工。
化学系梅佩礼教授规划，厥特夫妇匿名捐 10 万美元建造。
被誉为"国内仅见之建筑物"，建筑规模和实验设备皆为国内首屈一指。
曾用作实验室、科学演讲厅、陈列室等。

—

The Science Hall was started in 1921 and completed in April, 1922.
This building was planned by Prof. Fred C. Mabee, who served as chair of the Department of Chemistry
at the University of Shanghai, and built with a donation of US $100,000 anonymously from Mr. and Mrs.
Treat of Pasadena, California. At the time it was one of the most advanced and pioneering buildings in
China, with the largest building scale and state-of-the-art lab equipment.
It was used as a laboratory, a science lecture hall, a showroom and so on.

SCIENCE
HALL

这所巍峨的圣殿，藏着古往今来伟大的精灵。他们的使命，是为解答宇宙间的真理。

SCIENCE HALL

科学馆地上有四层。第一层为物理部，第二层为生物部，
第三层为化学部，第四层为地质部。地下一层用来贮藏危险物和制造汽水。
实验设备达到当时国内最先进水平，为理科进一步发展"提供了广阔的前景"。

—

The hall consists of four floors above the ground. The first floor houses the Department of Physics;
the second floor is home to the Department of Biology; the third floor contains the Department of Chemistry;
the fourth floor is the seat of the Department of Geology. Its ground floor is used for storing dangerous
chemicals and making aerated water. On its opening, our experimental equipment, the most advanced in China,
gave a bright future to the further development of basic sciences.

大礼堂与思魏堂

1936 年初举行破土礼，1937 年 5 月竣工。

大礼堂与思魏堂为连体建筑，建筑平面呈 L 形。

思魏堂位于大礼堂北侧，上层为礼拜堂，下层设办公室、教员休息室等。

建筑立面装饰哥特式十字花窗、尖券窗，外窗用细柱间成多个竖向小窗，垂直立体感强。

大礼堂为庆祝沪大建校 30 周年建造，思魏堂为纪念魏馥兰校长而建。

大礼堂与思魏堂也是学校标志性建筑。

—

The ground-breaking ceremony was held in 1936 and the Auditorium and White Chapel were erected in May 1937. The Auditorium and White Chapel are joint buildings, with a structure looking like "L" on a plane. The White Chapel is located on the north of the Auditorium. The upper floor was used for the Chapel, and the lower floor for offices, teachers' lounges, and other purposes.

The facade of the building is decorated with Gothic cross-shaped windows and pointed arch windows, and the external windows are divided into several small vertical windows with thin columns, which attract the eye upwards and give the edifice a strong sense of verticality. The Auditorium was built to celebrate the 30th anniversary of the University of Shanghai, while the White Chapel was built to commemorate President Francis Johnstone White. The Auditorium and White Chapel is also a landmark structure of the university.

Auditorium and White Chapel

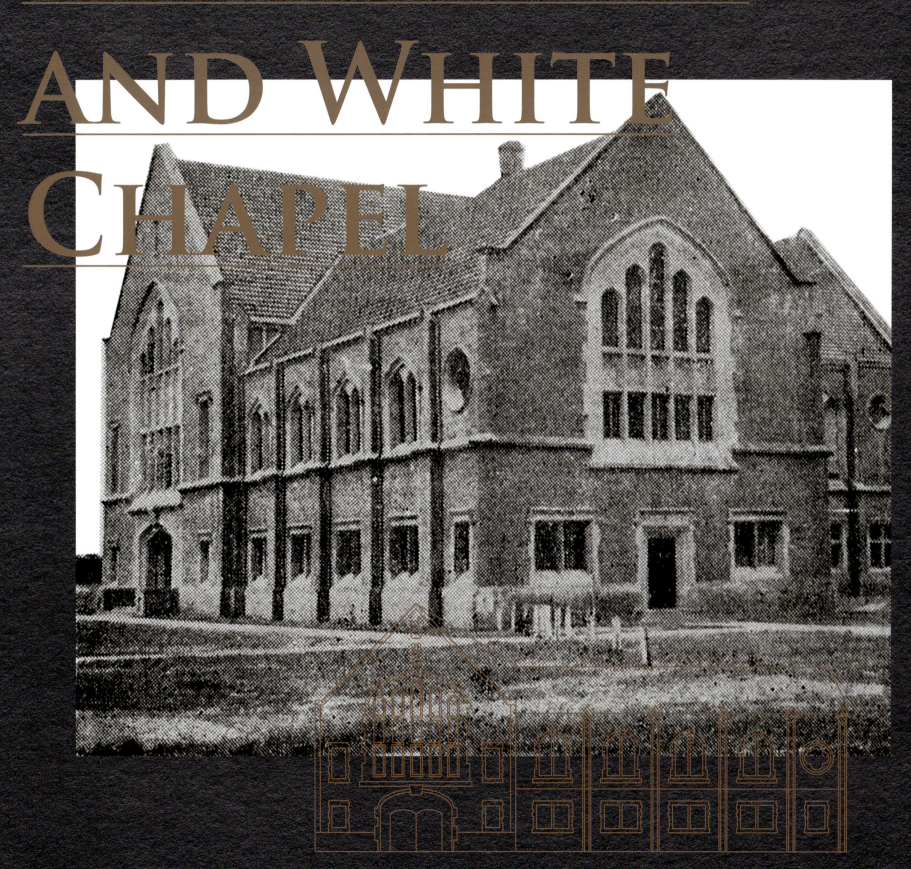

抗战期间，沪大校园被日军侵占，学校搬到位于公共租界的真光大楼继续办学。
抗战胜利后，1946 年 4 月 10 日，师生终于重返校园。凌宪扬校长在报告中写道：
"学生们不但加倍用功来表现他们精神振奋，而且不分男女，都以亲手清除碎砖烂瓦、
残根杂草来帮助重建，手上打了血泡与劳工们并肩干活。"
同年 11 月 23 日，在大礼堂与思魏堂隆重庆祝建校 40 周年，校友和嘉宾欢聚一堂。

—

During the War of Resistance against Japan, the university campus was occupied by Japanese invaders and
the University of Shanghai moved to the China Baptist Publication Building in the Shanghai International Settlement.
After China's victory in the war, the teachers and students eventually returned to the campus on April 10, 1946, just as
President Ling Xianyang wrote in his report, "To demonstrate that they are in high spirits, our industrious students, whether
boys or girls, spared no efforts to remove the broken bricks, tiny pieces of tiles and rooted out plant studs or weeds
with their own bare hands to help with the reconstruction of our campus. Even with blood blisters on their hands,
they still worked together with hired laborers." On November 23rd of the same year, alumni and guests gathered together to
celebrate the 40th anniversary of the founding of the university in the Auditorium and White Chapel.

AUDITORIUM AND WHITE CHAPEL

麦氏医院

1917 年建成，同年 5 月 16 日举行开业典礼。

建筑费用由中外人士募捐，麦克来氏捐款数额最高，故名麦氏医院，又称为普济医院。

第一位驻院医生雷盛休博士曾被孙中山授予勋章，学校慕名聘为校医。

徐志摩捐款 50 墨银并记录医院建造经过："本校谋建医院，济穷利众。美国善士捐来五千元，又向校友筹措千元。现既匀工从事，今冬可以落成。主院事者，雷医生也。"

—

Mcleish Infirmary was erected in 1917 and inaugurated on May 16 of the same year.

Both Chinese and foreigners donated the construction fee. The highest amount of donations came from Mr. McLeish, and for this reason it was named Mcleish Infirmary, and also came to be known as Puji Infirmary.

The first resident doctor in the infirmary was George Arthur Huntley, who was awarded the Order of Golden Grain by Sun Yat-sen. As a result, the College appointed him as College Physician.

Xu Zhimo donated 50 Mexican dollars and recorded the construction of the Infirmary as follows: "The University planned to build an infirmary to help the poor and benefit the public. An American philanthropist donated 5,000 Mexican dollars himself and raised 1,000 Mexican dollars from his fellow schoolmates. Now it is under construction and will be completed in the coming winter. Dr. George Arthur Huntley will take the charge of the infirmary."

McLeish
Infirmary

四维风月，潇然尘外。

体育馆

1916 年 12 月测地绘图，1918 年竣工，1919 年举行开馆典礼，是沪上各校第一座体育馆。

混合结构，屋顶陡峭，外墙壁柱逐层收缩，主入口尖券门洞线脚层层缩进。

外墙窗下装饰哥特式花纹，窗洞两侧装饰齿形线脚。

学校体育为华东地区翘楚，田径、体操各有所长，篮球、网球、足球和垒球等在校际比赛中多次夺冠。

—

Haskell Gymnasium, which was to become the first dedicated gym in universities in Shanghai at that time,
was surveyed and mapped in December 1916, completed in 1918, and inaugurated in 1919.
With a brick-wood structure, the building was designed with a steep roof and pilasters on exterior walls contracted layer
by layer. At the main entrance, characterized by a pointed arch, the door was inset among several layers of masonry.
Windows on the exterior walls were decorated with Gothic patterns, with both sides of the window opening having a design of
tooth-shaped moldings.The University of Shanghai ranked among the top institutions in East China for sports,
with a reputation for excellence in athletics and gymnastics and won championships many a time in basketball,
tennis, football and softball in interscholastic competitions.

HASKELL

GYMNASIUM

那儿呈给你的是：斯巴达的壮士，Apollo 的典型。

有谁说我们是病夫，我要领他去瞻仰瞻仰瞻仰我们这座古希腊之宫，

HASKELL GYMNASIUM

艾德蒙堂

1932 年建成。砖混结构，立面中部突出，上有小尖塔和雉堞。
艾德蒙夫人及女儿捐款 1.08 万美元，故名艾德蒙堂。曾用作女生体育馆（健身房），
内附女生膳堂和音乐室。

—

Built in 1932, Edmands Hall, constructed with bricks and concrete, consists of a salient central facade,
and a roof decorated by torrents and crenellation. It was named Edmands Hall because Mrs. Edmands and
her daughter donated 10,800 US dollars towards the construction of the building. The hall was formerly
used as a girls' gymnasium, with a girls' dining room and a music room inside.

EDMANDS

HALL

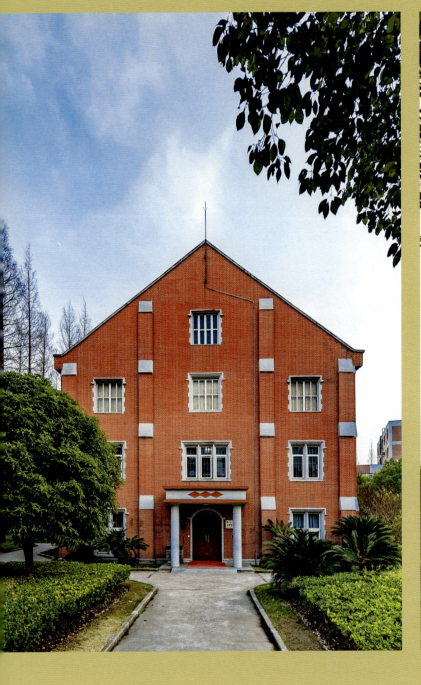

EDMANDS
HALL

音乐堂

1935 年 5 月 11 日行破土礼，年底建成。

砖混结构，屋顶陡峭，主入口为山墙门廊式，尖券窗用直棂分隔，北立面有三联拱尖券，

扶壁以小尖塔收头。曾用作中学部礼堂，后改作音乐堂。学校成立多个西洋和民族乐队，

有"远东第一小号"之称的音乐家朱起东曾任音乐系主任。

—

The ground-breaking ceremony was held on May 11, 1935 and the hall was built at the end of the year.

Built with bricks and concrete, it consists of a steep roof, with a gable porch as the main entry, windows separated

by mullions, a north wall decorated by 3 arch tips, and a buttress ending with a tower tip.

It was formerly used as Academy Assembly Hall, and later as an auditorium for the Department of Music of the

University of Shanghai. The university students set up various prestigious bands playing both Western and Chinese

music. Zhu Qidong, a musician credited as "the No. 1 Trumpet Player in the Far East",

served as chair of the Department of Music for a period of time.

MUSIC HALL

MUSIC HALL

军工路校区
JUNGONG ROAD CAMPUS

复兴路校区
FUXING ROAD CAMPUS

沪大培养学生"服务于社会和国家的崇高理想",并"使之愿意为这一理想去服务,并必要的话,去牺牲"。

—

The University of Shanghai fostered students with "the lofty idea of serving their society and country", "willing, if necessary, to make sacrifices".

—

创办中国第一个大学社会实验基地"沪东公社",学生以专业知识服务于杨树浦周边工厂和乡村。公社成为中国劳工教育和现代工人社区文化样板,享誉海内外。在辛亥革命、五四运动、五卅运动、抗日战争和解放战争中高举爱国主义旗帜,涌现出刘湛恩、李公朴等革命英烈和李一氓、涂长望等爱国英杰。

—

At the Yangtzepoo Social Center(currently Yangpu District), the first university center for social experiment in China, the students served the industries and villages around Yangtzepoo with their professional knowledge. In this way, the Center became the model for China's labor education and community culture of modern factory workers that came to enjoy a world-wide reputation. Students of the University of Shanghai bravely safeguarded the country during the Revolution of 1911, the May 4th Movement, the May 30th Movement (1925), the War of Resistance against Japan, and the War of Liberation, and revolutionary martyrs such as Liu Zhan'en and Li Gongpu, as well as other patriotic heroes such as Li Yimeng and Tu Changwang, all came to the fore.

—

学校吸取美国博雅教育理念，引导学生全面发展，培养出众
多知名英才：徐志摩、苏祖斐、纪育沣、邱式邦、冯亦代、
欧阳山尊、林乐义、李储文、丁景唐、胡壮麒、沈之荃、李
道豫、汪尔康等。

–

The University of Shanghai adopted the idea of America's Liberal Arts education, guiding students to develop in an all-around way, and cultivated lots of renowned figures in history. Its ethos has produced many celebrated talents, such as Xu Zhimo, Su Zufei, Ji Yufeng, Qiu Shibang, Feng Yidai, Ouyang Shanzun, Lin Leyi, Li Chuwen, Ding Jingtang, Hu Zhuangqi, Shen Zhiquan, Li Daoyu and Wang Erkang.

PART 2

思裴堂

1914 年暑假始建，1915 年 4 月竣工。

第二幢大建筑，砖混结构，双坡屋面，南立面门廊处竖立八根科林斯柱子，下饰宝瓶状栏杆。

第一层用作教室，其余为大学三四年级男生寝室，共 36 间，每间住 2 人。裴理克博士募捐建筑费用 2.2 万墨银，故名思裴堂。1967 年大修，改假三层坡顶为四层平顶。曾用作教室和男生宿舍。

—

Ground was broken for Breaker Hall in the summer of 1914 and it was completed by April 1915.

As the second largest building in the University of Shanghai, it has a brick-concrete structure, with a double slope roof, and 8 Corinthian columns in its south porch, under which there are bottle-shaped railings.

The first floor was used as classrooms, and the other floors were dormitories for male students in their third and fourth years of university, with 36 rooms, each housing 2 persons. It was named Breaker Hall in honor of Dr. Manley Juhan Breaker, who fundraised 22,000 Mexican dollars for the building. It underwent an overhaul and the sloping roof on the third story was changed into a flat roof on the fourth story in 1967. It was formerly used as classrooms and as a men's dormitory building.

BREAKER HALL

思伊堂

1919 年竣工，建筑费用约 6 万墨银。

砖木结构，屋顶陡峭，老虎窗林立，主入口处有两层凸窗，窗下墙壁装饰尖券图案，

走廊以连续四圆心券间隔。曾用作男生宿舍。1984 年大修，将原坡顶改为平顶。

—

The hall was erected in 1919, with a construction fee of 60,000 Mexican dollars.

With a brick-timber structure, it has a steep roof, a number of dormer windows, a two-layer bay window

above the main entrance, and a pointed arch pattern was decorated on the wall under the window.

The corridor is separated by a series of four-centered arches.

It was formerly used as a men's dormitory building, and repaired and changed the roof from slope to flat in 1984.

EVANSTON
HALL

连朝烟雾暗江边，
扫榻西窗听雨眠。
睡觉始知淫雨止，
一轮红日蔚蓝天。
喜晴瓦雀更相催，
早起纱窗四扇开。
雨后乾坤齐洗雾，
浦东如画入帘来。
——思伊堂

EVANSTON HALL

思孟堂

1920 年竣工，建筑费用约 6 万墨银。

1924 年命名为思孟堂。砖混结构，清水红砖墙，入口设在建筑中部，上方配凸窗，室内走道有尖券。

曾是附属中学下院，用作教学楼和宿舍。

—

The hall was erected in 1920, with a construction fee of 60,000 Mexican dollars.

It was named Melrose Hall in 1924. With a brick concrete structure, its walls are built with red bricks.

It has an entrance in the middle, with a bay window above it and a pointed arch in the corridor inside.

It was formerly used as the Junior Academy Building, with classrooms and dormitory rooms.

MELROSE HALL

学生描写思孟堂景色："青天凭窗可以看到江边，Love Lane 上的垂柳，
江畔的白云，静静地，没有喧哗，没有骚扰。黄昏的江岸，落日晚霞更会增加几分诗情画意。
明日初生，或布满繁星的晚间，留声机里常会放出 Waltz 轻快的旋律，
爵士音乐婉转低缓的歌声，同学们轻快的笑语，用功同学朗朗的书声。"

MELROSE
HALL

思雷堂

1922 年竣工。思雷堂与思孟堂的建筑费用一样，
建筑风格基本一致。1924 年命名为思雷堂。
曾是附属中学上院，用作中学三四年级教室、宿舍及中学部礼堂。

—

Erected in 1922, Richmond Hall, which shares a construction cost and style with Melrose Hall,
was named Richmond Hall in 1924. It was formerly used as Academy Senior Building with classrooms for
Grade 3 and Grade 4 students, as well as dormitory rooms and the Middle School Auditorium.

RICHMOND HALL

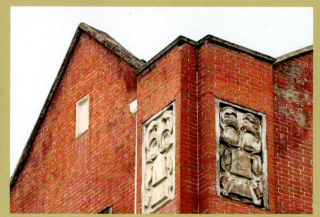

绿树荫中，弦歌不辍。

RICHMOND HALL

怀德堂

1922 年动工，1923 年秋竣工。
砖木结构，原屋顶陡峭设老虎窗，正立面中部有凸窗，建筑转角似扶壁，走廊有尖券洞。
厥特夫妇匿名捐助 10 万墨银建造，命名怀德堂。曾用作女生宿舍。

—

Ground was broken for Treat Hall in 1922 and its construction was completed in the fall of 1923.
With a brick-timber structure, the original hall has a steep roof with dormer windows and a bay window in the
middle of the facade. The corner of the hall looks like a buttress, and there is a pointed arch hole in the corridor.
It was named Treat Hall in honor of Mr. and Mrs. Treat, who donated 100,000 Mexican dollars to
the construction of the hall. It was formerly used as a women's dormitory building.

TREAT

HALL

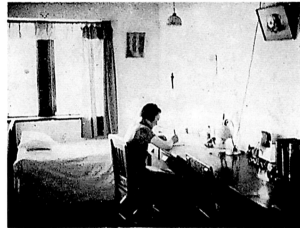

TREAT
HALL

沪大是国内最早招收女生的大学之一。1920 年开始招收女学生，
是年秋即有 4 名女生入学。怀德堂作为女生宿舍是校内一个独立区域，被视为"禁宫"，男生"休想得入"。
中国著名儿科专家、儿童营养学家、医学教育家苏祖斐，抗战时期外交亲历者、
联合国首批礼宾官严幼韵等人都曾住在怀德堂。

—

As one of the first universities to enroll female students in China, the University of Shanghai
began to welcome female students in 1920, and four girls entered the university in the fall of that year.
As an independent unit within the university, Treat Hall was regarded as a "forbidden area" as no male was allowed
to enter it. Here once dwelled such famous alumni as Su Zufei, a famous Chinese pediatrician, child nutritionist and
medical educator, and Yan Youyun, who later became a distinguished diplomat, a witness to the War of Resistance
against Japan and one of the first cohort of officers of the United Nations.

思福堂

1936 年募建，1937 年建成。
砖木结构，清水红砖墙，白色窗套，开窗方正，南立面有浮雕式圆拱，
阳台装饰卷草纹柱子，三角形山墙外露。
曾用作女教员宿舍。

—

Fundraising and design for the hall began in 1936 and it was erected in 1937.
With a brick-timber structure, it has red brick walls, white window covers, square window openings,
a south facade with embossed arches, a balcony decorated with pillars with scroll grass pattern,
and exposed gables. It was formerly used as the Female Faculty Residence.

VIRGINIA
HALL

馥赉堂

1947 年举行奠基礼，1948 年建成，1949 年启用，是沪大最晚的建筑。
砖混结构，假四层坡屋面，两个入口门厅，分别为三角山墙和雉堞墙样式。
白色仿石线窗套，底层窗洞为哥特式，二层以上简化为现代式。
曾用作大学女生宿舍。

—

The groundbreaking ceremony was held in 1947, and the hall was erected in 1948 and available to use
in 1949. It was the latest building of the period added to the University of Shanghai. With a brick-concrete
structure, it consists of a fake four-story sloping roof, and two entrance halls, which are in the style of a triangular
gable wall and a crenelated wall respectively. It has imitation white stone window covers,
and Gothic bottom window holes, while the second floor and above are in the simplified modern style.
The hall was formerly used as a girls' dormitory building.

Franklin-Ray Hall

FRANKLIN-
H

RAY
ALL

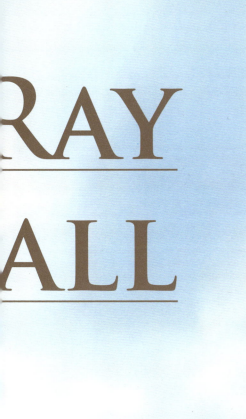

东
堂

1909 年建成。砖混结构，立面中部突出，上有小尖塔和雉堞。
初两层，楼上为膳厅，楼下为盥洗室。后扩建为三层。
徐志摩 1915 年进入沪大就读，居住在北堂，用餐和盥洗则在东堂。

—

East Hall was erected in 1909. With a brick-concrete structure,

its center of the facade is prominent with small minarets and rafters.

Originally, it was a two-story building with a dining room upstairs and facilities downstairs.

Later, the third floor was added. In 1915, Xu Zhimo enrolled in Shanghai College as a student.

At that time, he lived in the North Hall and had meals and washed up in East Hall.

EAST

HALL

徐志摩在校时，学习成绩优异，古典文学功底深厚，创作多篇诗文，
曾担任《天籁》汉文主笔和年级级长。1916年11月《天籁报》第4卷第3号刊载徐志摩的骈文《送魏校长归国序》：
去乡土万里，越重洋来异国，诲人不倦，毋惜辛苦。卷利己之心，抱救人之念。斯其德音之在人，不以深隃。况
于叔世颓风，道德淘淘，曾有狂澜莫挽之叹。乃有君子者，砺身作则，热血感人，前弊务尽，树德务滋，如吾魏
校长者，足以当之而无愧矣。先生温温长者，望之也威，接之也和。讷然如不能言，而情意恂挚，循循善诱，虽
不良亦已化矣。始先生未来是土，荒滨草原，浪涛溅渍，沙鸥海鸟，时复出没，星芒渔火，相与辉照。先生独劳
心焦虑，施意经营，数年之间，蓄然美备。广厦连诗，学子兴来，建始有方，守成兼理。所成者岂独辟荒陬除草
莱之功邪？拨盲心而涤污思，治璞冶金，括垢磨光，大德在人，可胜量哉！今先生且归国，稚鸟依母，不胜恋系
之忧。先生爱我，其将有以益吾后进未已也。

—

At Shanghai College, Xu Zhimo, whose academic performance was excellent,
demonstrated a profound knowledge of classical literature, and created a great number of poems and essays.
He was head of his grade and once served as chief writer in Chinese for *The Voice*. In November 1916, *The Voice*, Vol.4, No. 3,
published Xu Zhimo's parallel prose, entitled *Preface to Farewell to President White*.

EAST
HALL

水塔

1930 年建成。砖混结构，共五层，建筑面积 298 平方米。

水塔建成后，保障师生用上洁净的自来水。1937 年 8 月遭日军炮击，水塔上部留下弹孔。

—

With a brick-concrete structure, the Water Tower, erected in 1930, has five storeys,

covering a total area of 298 square meters. After the completion of the Water Tower, clean running water was available to

both teachers and students. Shelled by Japanese forces in August 1937, the tower had bullet holes in the upper part.

WATER
TOWER

水塔上的弹孔

the bullet hole in Water Tower

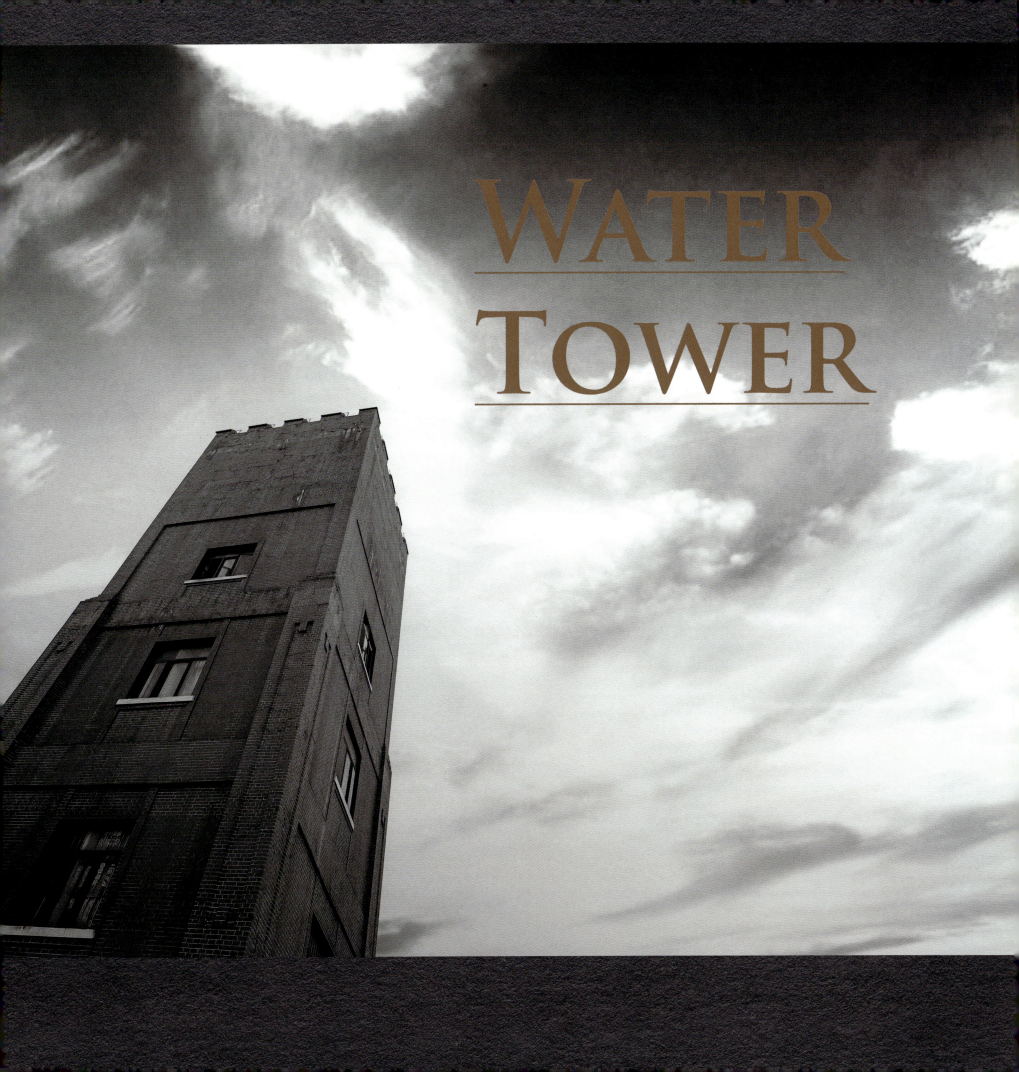

WATER
TOWER

大学膳堂

1931 年 4 月 24 日举行破土礼，同年底建成。
曾用作男生膳厅。

—

The ground-breaking ceremony was held on April 24, 1931 and the Dining Hall was
erected at the end of the year. It was formerly used as the Men's Dining Hall.

DINING HALL

军工路校区
JUNGONG ROAD CAMPUS

复兴路校区
FUXING ROAD CAMPUS

军工路校区现存14幢2-3层独立式小别墅，原为沪大教员住宅，其中13幢建于1915至1925年。这些建筑采用坡屋顶，设老虎窗，外墙用红砖或青、红两色砖混砌，采用预制混凝土花饰门窗套，窗多用花饰砖拱，室内设有砖砌壁炉。

—

There are currently 14 detached villas existent on Jungong Road Campus, each with 2 or 3 storeys, of which 13 were built from 1915 to 1925. They were formerly used as the residences for teachers of the University of Shanghai. These buildings use sloping roofs, dormers, and red bricks or red bricks mixed with indigo bricks for their outer walls, precast concrete door covers and window covers with flower patterns; the windows are mostly decorated with arched bricks in flower patterns, and there is a fireplace made with bricks inside each building.

—

其中7幢组成沪江国际文化园，现设德国、美国、英国、日本、澳洲、北欧、法国7个文化交流中心，成为独具魅力的校园国际文化社区。

—

Seven of the buildings constitute the Hujiang International Cultural Park, in which seven cultural exchange centers have been established for Germany, America, Britain, Japan, Australia, France and Nordic-Baltic countries respectively. They have now become a uniquely appealing international zone for cultural communities on campus.

—

沪大广聘天下名师，董景安、吕思勉、葛学溥、余日宣、余上沅、雷洁琼、蔡尚思、施蛰存、章靳以、徐中玉、谢希德等都曾在此任教。

—

The University of Shanghai appointed such world-renowned teachers as Dong Jingan, Lü Simian, Daniel H. Kulp, Yu Rixuan, Yu Shangyuan, Lei Jieqiong, Cai Shangsi, Shi Zhecun, Zhang Jinyi, Xu Zhongyu, and Xie Xide.

PART 3

教员住宅 103\104 号

1907 年竣工。1915 年春火灾后重新修缮，1916 年夏竣工。
初为化学系主任梅佩礼教授住宅，现为中德国际学院办公楼。

—

The residence was erected in 1907, rebuilt in the spring of 1915 after a destructive fire,
and was completed in the summer of 1916. Initially used as the residence of Prof. Fred Carleton Mabee,
chair of the Department of Chemistry, it currently houses the Sino-German College.

Faculty Residence No.103/ 104

教员住宅 203\204 号

1916 年建成。初为李佳思教授住宅，后用作沪大校长樊正康住宅，
机械学院时期为家属楼，现为美国文化交流中心。

—

It was erected in 1916, initially used as Prof. Charles Lucas Bromley's residence, later as the residence of Fan Zhengkang, the University's President. It served as a faculty residence during the period of Shanghai Mechanical College, and currently houses the American Culture Center.

FACULTY RESIDENCE NO.203/ 204

教员住宅 201\202 号

1917 年建成。初为雷盛休医生住宅，后用作韩森教授住宅。
机械学院时期为家属楼，现为德国文化交流中心。

—

It was erected in 1917, initially used as George Arthur Huntley's residence,
later as the residence of Prof. Han Sen. During the period of Shanghai Mechanical College it
served as a faculty residence, and currently houses the German Culture Center.

FACULTY RESIDENCE

No.201/ 202

教员住宅 205 号

1920 年建成。初为赫齐佳教授住宅。
机械学院时期为家属楼，现为英国文化交流中心。

—

It was erected in 1920, initially used as Prof. Henry Huizinga's residence,
later as a faculty accommodation during the period of Shanghai Mechanical College.
It currently houses the British Culture Center.

Faculty Residence No. 205

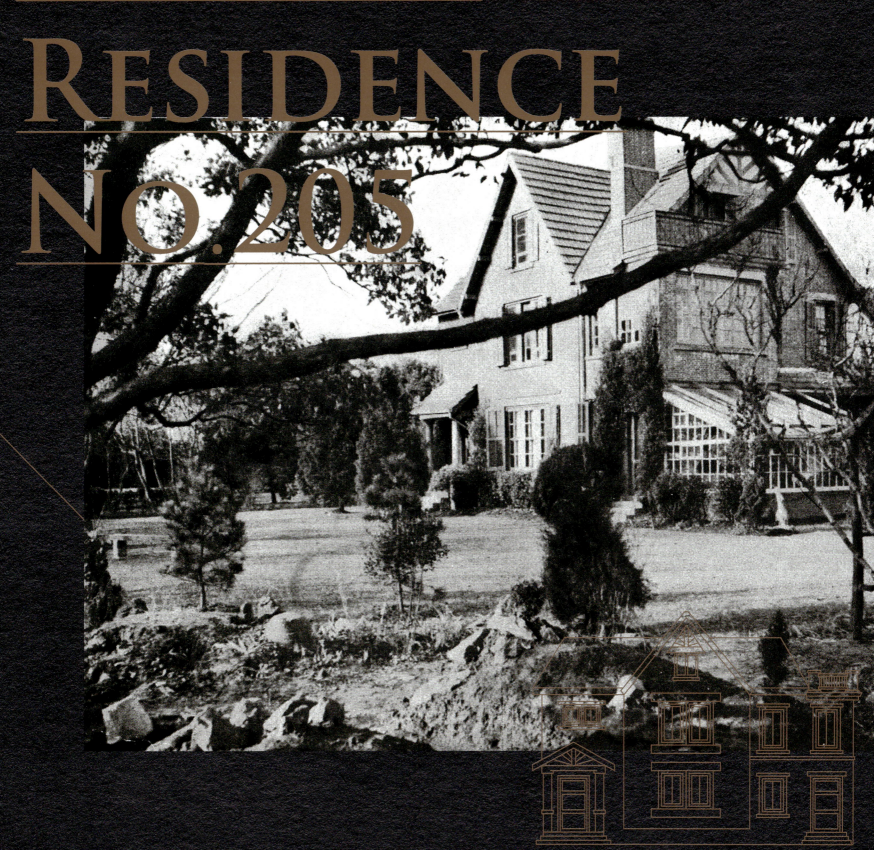

教员住宅 206 号

1921 年建成。原为教员住宅，
机械学院时期为家属楼，现为北欧文化交流中心。

—

It was erected in 1921 and used as a faculty residence, then a residential building during the period of
Shanghai Mechanical College. It currently houses the Nordic-Baltic Culture Center.

FACULTY RESIDENCE NO.206

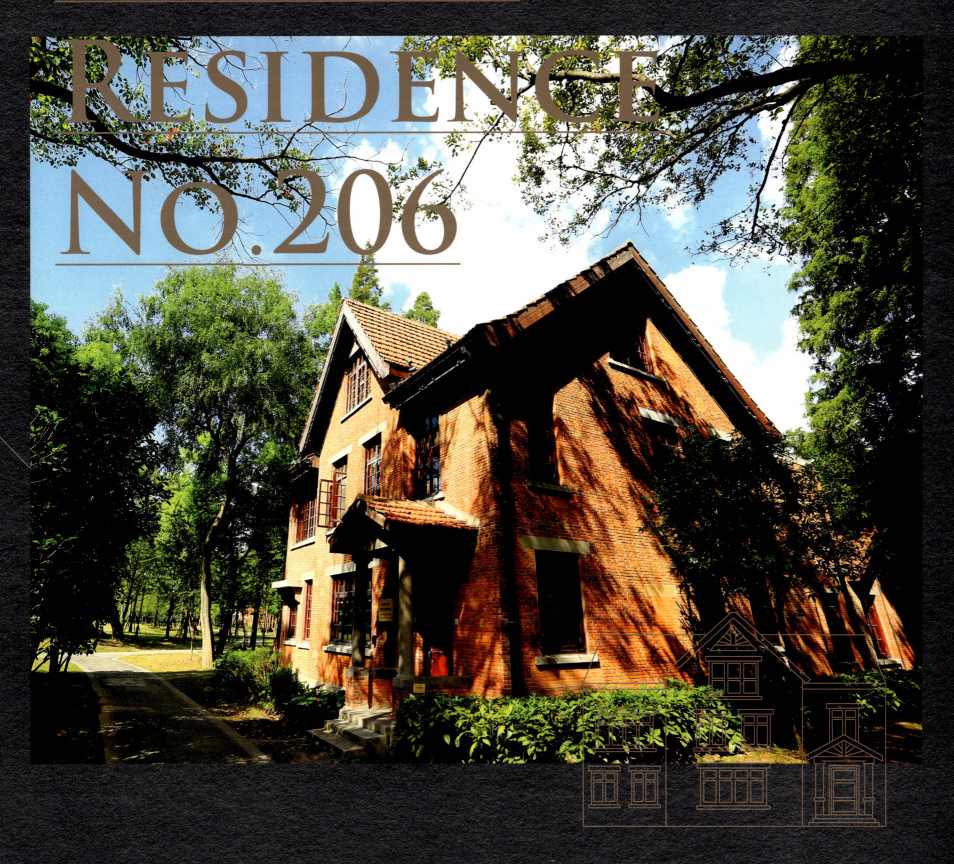

教员住宅 207 号

1921 年建成。原为教员住宅，
机械学院时期为家属楼，现为澳洲文化交流中心。

—

It was erected in 1921 and used as a faculty residence,
then a residential building during the period of Shanghai Mechanical College.
It currently houses the Australian Culture Center.

Faculty Residence No. 207

教员住宅 108 号

1922 年建成。原为刘湛恩校长的住宅，现为刘湛恩烈士故居。

刘湛恩是中国近代著名爱国教育家和社会活动家，先后获东吴大学理学学士、芝加哥大学教育学硕士、哥伦比亚大学哲学博士学位。1928 年出任沪大校长，是当时中国最年轻的大学校长之一。

—

It was erected in 1922, initially used as the residence of President Liu Zhan'en (Herman C. E. Liu), currently preserved as the Former Residence of Martyr Liu Zhan'en.

As a famous patriotic educator and social activist in modern China, Dr. Liu Zhan'en obtained a bachelor's degree in science from Soochow University (SCU), a master's degree in education from the University of Chicago, and a doctor's degree in philosophy from Columbia University. He was appointed as president of the University of Shanghai in 1928 and was one of the youngest university leaders in China at that time.

Faculty Residence No.108

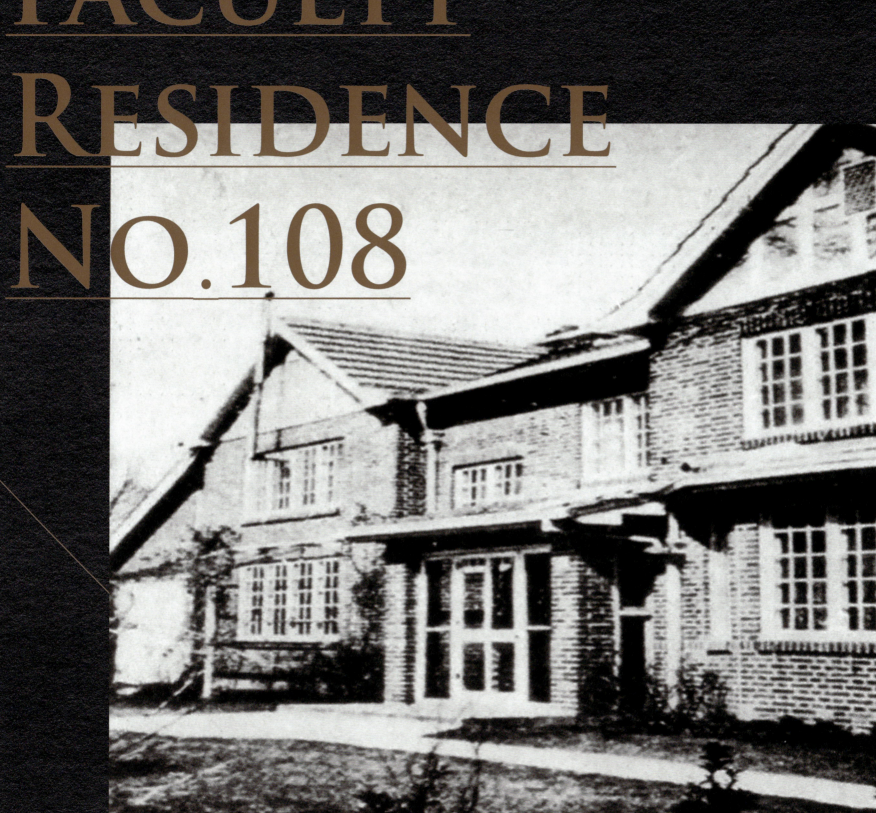

1928 年 2 月 25 日，沪大为刘湛恩举行就职典礼。
大学院院长蔡元培、工商部部长孔祥熙、上海市市长张定璠、淞沪警备司令熊式辉、
圣约翰大学校长卜舫济、复旦大学校长李登辉等出席。

—

On February 25th, 1928, an inauguration ceremony for President Liu Zhan'en was held at the University
of Shanghai. Present at the ceremony were many leading figures from all walks of life, including Cai Yuanpei,
President of the Grand College (roughly equivalent to the National Council of Education); Kong Xiangxi,
Minister of Industry and Commerce; Zhang Dingfan, Mayor of Shanghai; Xiong Shihui,
the Commanding Officer of the Shanghai Garrison; Francis Lister Hawks Pott, President of St. John's
University and Li Denghui, President of Fudan University.

刘湛恩就任校长后，对沪大进行"中国化"校政改革，广聘知名学者任教，
开设城中区商学院，沪大的办学规模和社会影响力迅速提升。

—

After Liu Zhan'en took office as president, he carried out the university's Sinicization reform in administration,
appointed lots of renowned scholars as teachers, and established the School of Business in the city center proper.
Under Dr. Liu's leadership, the University of Shanghai enjoyed a rapid growth both in scale and social influence.

抗战爆发后，刘湛恩被推举为上海各界人民救亡协会理事、上海各大学抗日联合会负责人
和中华全国基督教协进会难民救济委员会主席，成为上海教育界抗日救亡领袖。
由于刘湛恩校长的抗日救亡言行，尤其是断然拒绝出任伪政府教育部长，被日寇列入暗杀名单。1938 年 4 月 7 日，
遭日伪特务刺杀，不幸当场殉难，是抗战中唯一牺牲的大学校长。1985 年被追认为"抗日革命烈士"。

—

When the War of Resistance against Japan broke out in 1937, Liu Zhan'en was recommended to serve as a director of Shanghai People's Salvation Society, head of the Anti-Japanese Federation of Shanghai Universities and chairman of the Refugee Relief Association Committee of the Chinese Christian Council. In short, he became the leader in Shanghai educational circles of the fight against the Japanese occupation. Because of his brave words and deeds in the Movement of Resistance against Japan to Save the Nation from Going Extinct, and especially his flat refusal to serve as the Minister of Education for the Japanese puppet government, President Liu was included in the assassination list drawn up by the Japanese government. On April 7, 1938 he was assassinated by a spy from the Japanese invaders and unfortunately died an instant death on the spot. As the only university president who died during the War of Resistance against Japan, he was posthumously recognized as a "revolutionary martyr fighting against the Japanese invaders" in 1985.

教员住宅 208 号

1923 年建成。

原为教员住宅，机械学院时期为家属楼，现为日本文化交流中心。

—

It was erected in 1923 and used as a faculty residence,

then a residential building during the period of Shanghai Mechanical College.

It currently houses the Japanese Culture Center.

Faculty Residence No.208

1936 年建成。

初为教师湛罗弼夫人为自己建造的住宅，后捐给学校。

文学家章靳以曾于 1951 至 1953 年间在此居住。

机械学院时期为家属楼，现为法国文化交流中心。

—

It was erected in 1936.

Mrs. R. E. Chambers built it as her own residence when teaching in the University of Shanghai.

After leaving China, she donated it to the university.Zhang Jinyi, a famous man of letters,

lived here from 1951 to 1953. Later it was used as a faculty residence during the period of Shanghai

Mechanical College. It currently houses the French Culture Center.

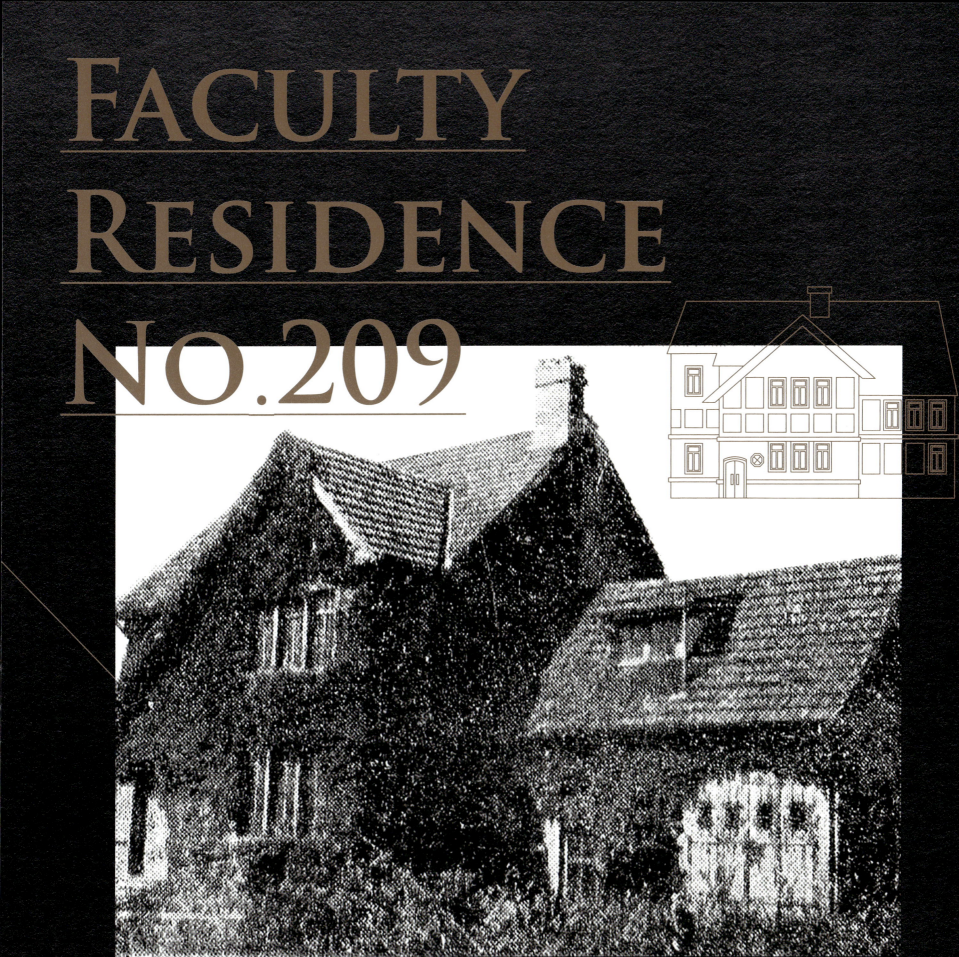

Faculty Residence

No.209

军工路校区
JUNGONG ROAD CAMPUS

复兴路校区
FUXING ROAD CAMPUS

1907 年，清廷诏谕振兴实业。1908 年，德文医工学堂购得城区法租界中心地块，以大力发展工科。其后八年，兴建德文科讲堂、工科讲堂等多幢普鲁士风格建筑。

—

In 1907, the Qing government gave the order to revitalize Chinese industries. In 1908, the Deutsche Medizinschule purchased a central plot of the French Concession, in the central urban area of Shanghai, with the intention of developing engineering significantly. Eight years later, it established various Prussian-style buildings, including the German Teaching Hall and the Engineering Courses Hall.

—

20 世纪初，引进德国工程教育制度，以"教授高深学术，养成医工专门人才"，实行世界先进的工科教育理念。20 世纪 30 年代，又汲取法国工程教育经验，培养"高深工业人才"。

—

In the early 20th century, the German-style educational system in engineering was introduced to implement the world's advanced engineering educational idea by "teaching advanced academic knowledge and cultivating first-class talents in medicine and engineering". In the 1930s, the university also learned from the experience of France's engineering education, again to foster "advanced talents in industry".

—

培养了钱令希、田炳耕、周立伟、闵乃本等一大批著名的科学家、教育家和工程技术专家。从工业救国、工业立国到工业强国，培养了一代代国家亟需的工业建设人才，为中国工业化发展做出重要贡献。

—

As a result, the University has produced a large number of well-known scientists, educators and experts in engineering and technology, such as Qian Lingxi, Tian Binggeng, Zhou Liwei and Min Naiben. From "saving the country by industry" to "building the country by industry" to "strengthening the country by industry", the University has nurtured generations of industrial construction talents urgently needed by the country, making great contributions to the industrialization of China.

PART 4

工科讲堂

1912 年开始建筑设计，1914 年落成。

德文医工学堂时期为工科讲堂，中法国立工学院时期为办公和教学楼。

国立上海高级机械职业学校时期为教学楼。

上海机械高等专科学校时期为图书馆。

—

Designed in 1912, the Hall was erected in 1914.

It has been successively used as the Engineering Courses Hall of the Deutsche Medizinschule, the office
and teaching building of the Franco Chinois Institute de Technique, the teaching building of Shanghai
National Senior Professional School of Machinery, and the Library of Shanghai Senior Mechanical School.

Engineering Courses

Hall

工科讲堂建筑图纸由倍高洋行的德国建筑师卡尔·培台克绘制，
建筑模型由工厂监督制定。砖混结构，德国普鲁士皇家机械学校风格，建筑面积 2072 平方米。
平面呈三面围合状，中部主入口处一、二层装饰塔司干柱，顶部为大尺度德国式圆弧形老虎窗。

—

The architectural drawings of the Engineering Courses Hall were made by Karl Baedecker,
a German architect from Becker & Baedecker. The building model was customized and designed under
the factory's experts' close supervision. Built in a brick-concrete structure, the hall covers a construction area of 2,072
square meters, following the building style of the Royal Prussian Mechanical School in Germany.
Enclosed on three sides, it is decorated with Tuscan columns on the first and second floors in the middle of the main
entrance while the top is fitted with large, arc-scale German dormer windows.

钟楼

医预科讲堂

德文科讲堂

德文科讲堂，1909 年建成，砖木结构。

医预科讲堂，1911 年建成，砖木结构。

钟楼，1909 年建成，砖混结构，象征"时不待人"的精神。

—

The German Teaching Hall was erected in 1909 in a brick-timber structure.

The Medicine Preparatory Hall was erected in 1911 in a brick-timber structure.

Erected in 1909 in a brick-concrete structure, the Clock Tower stands

for the ethos of the university that time and tide wait for no man.

GERMAN
TEACHING
HALL

Medicine
Preparatory
Hall

1910. Schule, Alumnat I.

CLOCK TOWER

学生宿舍

第二宿舍，1911 年建成。第三宿舍，1914 年建成。
第四宿舍，1915 年始建，1917 年落成。三幢宿舍均为砖木结构。

德文医工学堂由倍高洋行设计，构思和布局以普鲁士皇家机械学校的设计方案为蓝本。
德国科隆国立联合机械学校首席教师伯恩哈德·贝伦子具体负责筹建工学堂。
1908 年开始动工，1917 年初第四宿舍完工。整个建设工程前后长达 8 年，校园占地面积 47.5 亩。

—

The Second Dormitory was erected in 1911. The Third Dormitory was erected in 1914.
The Fourth Dormitory was started in 1915 and completed in 1917. All of the three dormitories
are brick-timber buildings.

Deutsche Medizinschule was designed by Becker & Baedecker, an architectural agency in Germany that took
the Royal Prussian School of Mechanical Technology as the blueprint in terms of design concept and layout.
Bernhard Berrens, chief lecturer of the Cologne National United Mechanical School of Germany,
was in charge of the preparatory work to construct the hall. Started in 1908, the whole project, covering a
total area of 47.5 mu(about 31,666.67 square meters) on the campus, took as long as 8 years to finish, with
the Fourth Dormitory completed at the beginning of 1917.

STUDENT
DORMITORIES

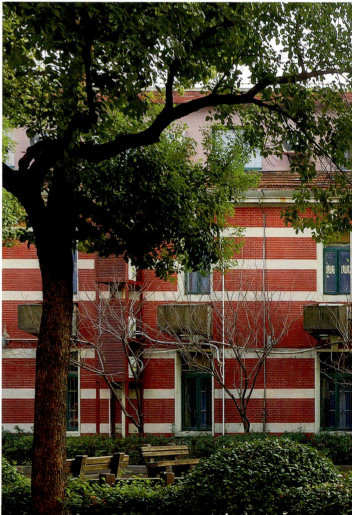

密氏校门

1916年建成，同年5月30日校门和门房正式启用。

为纪念沪大才华横溢、英年早逝的米拉德教授，故名密氏校门。

1953年底校门移至军工路旁，1992年建校86周年时按原样1.23倍放大改建。

—

Erected in 1916 and completed on May 30 of the same year, the Gate was named after William Harold Millard, the talented former professor who died young while serving as a teacher at the university. It was relocated to Jungong Road at the end of 1953, rebuilt and enlarged to 1.23 times the size of the original gate in 1992.

军工路校区历史建筑
ARCHITECTURAL HERITAGE OF JUNGONG ROAD CAMPUS

建筑原名 Former Name	竣工时间 Time for Completion	现在用途 Now Use	保护类别 Protection Category
北堂 North Hall	1907	已拆除 Demolished	/
教员住宅 101/102 号 Faculty Residence No.101/102	1907	已拆除 Demolished	/
教员住宅 103/104 号 Faculty Residence No.103/104	1907	中德国际学院办公楼 Office Building of Sino-German College	优秀历史建筑 Excellent Architectural Heritage
教员住宅 105 号 Faculty Residence No.105	1908	已拆除 Demolished	/
教员住宅 106/107 号 Faculty Residence No.106/107	1908	已拆除 Demolished	/
思晏堂 Yates Hall	1908	党校办办公楼 Office Building of the USST Party Committee and the President	优秀历史建筑 Excellent Architectural Heritage
东堂 East Hall	1909	暖屋超市 Nuanwu Store	历史建筑 Architectural Heritage
思马堂 Melrose Hall	1913	已拆除 Demolished	/
思裴堂 Breaker Hall	1915	第三学生宿舍 The Third Student Dormitory	优秀历史建筑 Excellent Architectural Heritage
密氏校门 Millard Gate	1916	军工路 516 号校门 Gate of No.516, Jungong Road	历史建筑 Architectural Heritage
教员住宅 203/204 号 Faculty Residence No.203/204	1916	美国文化交流中心 American Culture Center	优秀历史建筑 Excellent Architectural Heritage
教员住宅 201/202 号 Faculty Residence No.201/202	1917	德国文化交流中心 German Culture Center	优秀历史建筑 Excellent Architectural Heritage
麦氏医院 Mcleish Infirmary	1917	沪江美术馆 Shanghai Art Gallery	优秀历史建筑 Excellent Architectural Heritage
体育馆 Haskell Gymnasium	1918	学生活动中心 Student Activity Center	优秀历史建筑 Excellent Architectural Heritage

建筑原名 Former Name	竣工时间 Time for Completion	现在用途 Now Use	保护类别 Protection Category
思伊堂 Evanston Hall	1919	第四学生宿舍 The Fourth Student Dormitory	优秀历史建筑 Excellent Architectural Heritage
教员住宅 205 号 Faculty Residence No.205	1920	英国文化交流中心 British Culture Center	优秀历史建筑 Excellent Architectural Heritage
教员住宅 109 号 Faculty Residence No.109	1920	家属楼 109 号 Faculty Residence No.109	优秀历史建筑 Excellent Architectural Heritage
思孟堂 Melrose Hall	1920	能动学院办公楼 Office Building of Energy and Power Engineering School	优秀历史建筑 Excellent Architectural Heritage
发电房 Power House	1921	已拆除 Demolished	/
教员住宅 206 号 Faculty Residence No.206	1921	北欧文化交流中心 Nordic-Baltic Culture Center	优秀历史建筑 Excellent Architectural Heritage
教员住宅 207 号 Faculty Residence No.207	1921	澳洲文化交流中心 Australian Culture Center	优秀历史建筑 Excellent Architectural Heritage
科学馆 Science Hall	1922	格致堂 Administration Building	优秀历史建筑 Excellent Architectural Heritage
思雷堂 Richmond Hall	1922	能动学院办公楼 Office Building of Energy and Power Engineering School	优秀历史建筑 Excellent Architectural Heritage
教员住宅 108 号 Faculty Residence No.108	1922	刘湛恩烈士故居 Former Residence of Martyr Liu zhan'en	优秀历史建筑 Excellent Architectural Heritage
思魏池 White Pool	1922	已拆除 Demolished	/
怀德堂 Treat Hall	1923	第五学生宿舍 The Fifth Student Dormitory	优秀历史建筑 Excellent Architectural Heritage
教员住宅 208 号 Faculty Residence No.208	1923	日本文化交流中心 Japanese Culture Center	优秀历史建筑 Excellent Architectural Heritage

建筑原名 Former Name	竣工时间 Time for Completion	现在用途 Now Use	保护类别 Protection Category
教员住宅 110、111、112 号 Faculty Residence No.110,111,112	1924	家属楼 110、111、112 号 Faculty Residence No.110, 111, 112	优秀历史建筑 Excellent Architectural Heritage
思乔堂 Georgia Hall	1924	已拆除 Demolished	/
教员住宅 113/114 号 Faculty Residence No.113/114	1925	家属楼 113-114 号 Faculty Residence No.113-114	优秀历史建筑 Excellent Architectural Heritage
图书馆 Library	1928 初建 1948 扩建	公共服务中心 Public Service Center	优秀历史建筑 Excellent Architectural Heritage
教员住宅 121-126 号 Faculty Residence No.121-126	1929	家属楼 121-126 号 Faculty Residence No.121-126	历史建筑 Architectural Heritage
教员住宅 115-120 Faculty Residence No.115-120	1930	家属楼 115-120 号 Faculty Residence No.115-120	历史建筑 Architectural Heritage
水塔 Water Tower	1930	勤工助学中心 Work-Study Center	历史建筑 Architectural Heritage
大学膳堂 Dining Hall	1931	教工之家 Faculty Activity Center	历史建筑 Architectural Heritage
艾德蒙堂 Edmands Hall	1932	体育部办公楼 Sports Teaching Department	优秀历史建筑 Excellent Architectural Heritage
音乐堂 Music Hall	1935	音乐系办公楼 Music Hall	优秀历史建筑 Excellent Architectural Heritage
教员住宅 209 号 Faculty Residence No.209	1936	法国文化交流中心 French Culture Center	优秀历史建筑 Excellent Architectural Heritage
思福堂 Virginia Hall	1937	国际交流处办公楼 International Affairs	优秀历史建筑 Excellent Architectural Heritage
大礼堂与思魏堂 Auditorium and White Chapel	1937	大礼堂 Auditorium	优秀历史建筑 Excellent Architectural Heritage
教员住宅 210 号 Faculty Residence No.210	1947	家属楼 210 号 Faculty Residence No.210	优秀历史建筑 Excellent Architectural Heritage
馥赉堂 Franklin-Ray Hall	1948	留学生公寓 International House	优秀历史建筑 Excellent Architectural Heritage

复兴路校区历史建筑
ARCHITECTURAL HERITAGE OF FUXING ROAD CAMPUS

建筑原名 Former Name	竣工时间 Time for Completion	现在用途 Now Use	保护类别 Protection Category
德文科讲堂 German Teaching Hall	1909	行政办公 Administration Building	历史建筑 Architectural Heritage
钟楼 Clock Tower	1909	校史馆 History Museum	历史建筑 Architectural Heritage
第一宿舍 The First Dormitory	1909	已拆除 Demolished	/
医预科讲堂 Medicine Preparatory Hall	1911	复兴路校区管委会办公楼 Management Committee Office of Fuxing Road Campus of USST	历史建筑 Architectural Heritage
第二宿舍 The Second Dormitory	1911	第二学生宿舍 The Second Dormitory	历史建筑 Architectural Heritage
工科讲堂 Engineering Courses Hall	1914	图书馆 Library	优秀历史建筑 Excellent Architectural Heritage
第三宿舍 The Third Dormitory	1914	第三学生宿舍 The Third Dormitory	历史建筑 Architectural Heritage
第四宿舍 The Fourth Dormitory	1917	第四宿舍 The Fourth Dormitory	历史建筑 Architectural Heritage

---

[1] 军工路校区历史建筑中，图书馆、大礼堂与思魏堂为上海市第二批优秀历史建筑，其余均为第四批优秀历史建筑。

Among the architectural heritages on Jungong Road Campus, the Library, Auditorium and White Chapel are in the second batch of excellent architectural heritages in Shanghai while the other buildings are in the fourth batch.

[2] 复兴路校区历史建筑中，工科讲堂为上海市第三批优秀历史建筑。

Among the architectural heritages on Fuxing Road Campus, the Engineering Courses Hall belongs to the third batch of excellent architectural heritages in Shanghai.

《悦读：百年上理优秀历史保护建筑》由上海理工大学党委书记吴坚勇、校长丁晓东主持组织编写，并得到学校相关部门和社会各界的关心支持。

校党委副书记孙跃东和孙红、叶磊、刘彬等对画册给予指导，校档案管理与校史研究顾问伍贻文、张忠赓、何建中、白苏娣、吴元骉、张德明、杨佐平多次关心，党校办、组织部、宣传部、财务处、后勤处、基建处、校友处等部门给予支持。

校档案馆孔娜、刘淑娟、吴禹星考证史料、撰写文稿，孔娜审校中文书稿并统筹各项工作；刘淑娟、张柳承担英文翻译，张顺生、Phil Williams 审校英文书稿；陆赉设计画册；张德明共享相关资料；胡鸣强绘制建筑原貌；王平、王博拍摄建筑图片；靳海进、时瑞、许振哲协助相关工作。

分管档案工作的副校长张华审阅了全书。上海书画出版社第五编辑室主任孙晖促成画册顺利出版。

在本书编辑出版和"建筑可阅读"项目中提供支持的领导、部门和个人还有许多，恕难一一列举，谨致以最诚挚的谢意。

上海理工大学档案馆

2021 年 7 月

AFTERWORD

后记

Wu Jianyong, Party Committee Secretary of USST, and Ding Xiaodong, President of USST were in charge of the compilation of the album *A Visual Feast: Architectural Heritage of Centennial USST*, which received the care and support of relevant departments of the university and people from all walks of life.

Sun Yuedong, Deputy Party Committee Secretary of USST, Sun Hong, Ye Lei and Liu Bin have given instructions and advice. The experts like Wu Yiwen, Zhang Zhonggeng, He Jianzhong, Bai Sudi, Wu Yuanbiao, Zhang Deming and Yang Zuoping, have also given guidance and offered instructions on several matters. Functional departments including the Office of the Party Committee and the Office of the President, Organization Department, Publicity Department, Finance Office, Logistics Department, Infrastructure Department and Alumni Office have provided great support.

Kong Na, Liu Shujuan and Wu Yuxing from USST Achives verified historical files and provided the Chinese version, Kong Na proofread the Chinese version; Liu Shujuan and Zhang Liu provided the English version, Zhang Shunsheng and Phil Williams proofread the English version; Lu Lai designed the album; Zhang Deming provided relevant historical files; Hu Mingqiang drew the buildings as they originally appeared; Wang Ping and Wang Bo took photos of the buildings; Jin Haijin, Shi Rui and Xu Zhenzhe also assisted in related work.

Zhang Hua, Vice President in charge of the Archives, has proofread the whole album. Sun Hui, Director of the Fifth Editorial Office of Shanghai Fine Arts Publisher, facilitated the publication of the album in its present form.

We would like to express our sincerest gratitude to all the leaders, departments and individuals that have offered support for the publication of this album and the project of "Readable Buildings".

USST Archives

July 2021

AFTERWORD

后记

图书在版编目（CIP）数据

悦读：百年上理优秀历史保护建筑 ：汉英对照 /
吴坚勇，丁晓东主编. -- 上海：上海书画出版社，
2021.9
ISBN 978-7-5479-2696-3

Ⅰ. ①悦… Ⅱ. ①吴… ②丁… Ⅲ. ①高等学校－古
建筑－教育建筑－上海－图集 Ⅳ. ①TU244.3-64

中国版本图书馆CIP数据核字(2021)第154563号

悦读：百年上理优秀历史保护建筑

吴坚勇　　丁晓东　主编

责任编辑	孙　晖　凌云之君
审　　读	田松青
装帧设计	陆　赟
技术编辑	顾　杰

出版发行	上海世纪出版集团 上海书画出版社
地　　址	上海市闵行区号景路 159 弄 A 座 4 楼
邮　　编	201101
网　　址	www.ewen.co www.shshuhua.com
E-mail	shcpph@163.com
制版印刷	上海雅昌艺术印刷有限公司
开　　本	889×1194　1/12
印　　张	12 $^2/_3$
版　　次	2021 年 10 月第 1 版　2021 年 10 月第 1 次印刷

书　　号	ISBN 978-7-5479-2696-3
定　　价	380.00 元

若有印刷、装订质量问题，请与承印厂联系